Foundation

transition

practice

for **AQA, Edexcel** and **OCR two-tier GCSE mathematics**

CAMBRIDGE
UNIVERSITY PRESS

The School Mathematics Project

Writing and editing for this edition   John Ling, Paul Scruton, Susan Shilton, Heather West
SMP design and administration   Melanie Bull, Pam Keetch, Nicky Lake, Cathy Syred, Ann White

The following people contributed to the original edition of SMP Interact for GCSE.

Benjamin Alldred	David Cassell	Spencer Instone	Susan Shilton
Juliette Baldwin	Ian Edney	Pamela Leon	Caroline Starkey
Simon Baxter	Stephen Feller	John Ling	Liz Stewart
Gill Beeney	Rosemary Flower	Carole Martin	Biff Vernon
Roger Beeney	John Gardiner	Lorna Mulhern	Jo Waddingham
Roger Bentote	Colin Goldsmith	Mary Pardoe	Nigel Webb
Sue Briggs	Bob Hartman	Paul Scruton	Heather West

CAMBRIDGE UNIVERSITY PRESS
Cambridge, New York, Melbourne, Madrid, Cape Town, Singapore, São Paulo, Delhi

Cambridge University Press
The Edinburgh Building, Cambridge CB2 8RU, UK

www.cambridge.org
Information on this title: www.cambridge.org/9780521690027

© The School Mathematics Project 2007

First published 2007

Printed in the United Kingdom at the University Press, Cambridge

A catalogue record for this publication is available from the British Library

ISBN 978-0-521-69002-7 paperback

Typesetting and technical illustrations by The School Mathematics Project
Other illustrations by Chris Evans and David Parkins
Cover design by Angela Ashton
Cover image by Jim Wehtje/Photodisc Green/Getty Images

Using this booklet

This booklet provides accessible, well graded exercises on topics in the Foundation tier up to the level of GCSE grade F. The exercises can be used for homework, consolidation work in class or revision. They follow the chapters and sections of the *Foundation transition* students' book, so where that text is used for teaching, the planning of homework or extra practice is easy.

Even when some other teaching text is used, this booklet's varied and thorough material is ideal for extra practice. The section headings – set out in the detailed contents list on the next few pages – clearly describe the GCSE topics covered and can be related to all boards' linear and major modular specifications by using the cross-references that can be downloaded as Excel files from **www.smpmaths.org.uk**

It is sometimes appropriate to have a single practice exercise that covers two sections within a *Foundation transition* chapter. Such sections are bracketed together in this booklet's contents list.

Any section in the students' book that does not have corresponding practice in the practice booklet is shown ghosted in the contents list.

To help users identify material that can be omitted by some students – or just dipped into for revision or to check competence – sections estimated to be at national curriculum level 3 or 4 are marked as such in the contents list and as they occur in the booklet.

Marked with a red page edge at intervals through the booklet are sections of mixed practice on previous work; these are in corresponding positions to the reviews in the students' book.

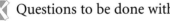 Questions to be done without a calculator are marked with this symbol.

Questions marked with a star are more challenging.

Answers to this booklet are downloadable from **www.smpmaths.org.uk** in PDF format.

Contents

continues >

1 Odds, evens, multiples, factors

| **A Odd and even numbers** | level 3 |
| **B Divisibility** | level 3 |

1 (a) Which numbers in the loop are even?

(b) Which are odd?

(c) Which are divisible by 10?

(d) Which are divisible by 4?

2 Each of these cards shows a letter and a number.

N	E	X	T	A	B	R	I	C	K
11	12	20	25	35	6	7	15	14	24

(a) Write down each letter that is on a card with

(i) an odd number   (ii) a number that is divisible by 5

(iii) a number that is divisible by 3   (iv) a number that is divisible by 7

(b) Rearrange each set of letters to give a means of transport.

| **C Multiples** | level 4 |
| **D Factors** | level 4 |

1 List all the multiples of 6 that are less than 40.

2 Which number in this list is not a factor of 15?

    5, 1, 6, 3, 15

3 Which numbers in this list are multiples of 9?

    6, 18, 27, 69, 54, 90

4 (a) Which numbers in the loop are multiples of

(i) 4   (ii) 5

(b) Which numbers in the loop are factors of

(i) 24   (ii) 30

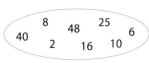

5 Use the clues to find the numbers.

(a)
| Between 5 and 10 |
| Factor of 16 |

(b)
| Between 40 and 50 |
| Multiple of 9 |

(c)
| Between 5 and 10 |
| Odd factor of 42 |

2 Mental methods 1

A Place value
level 4

1 In the number 372 984, the figure 9 stands for 9 hundreds or 900.

   (a) What does the figure 2 stand for?

   (b) What does the figure 7 stand for?

2 Write two hundred and fifty thousand in figures.

3 Work these out in your head.

   (a) 2615 + 100   (b) 33 961 − 1000   (c) 12 574 + 300   (d) 972 513 + 2000

4 The mileometer on Sarah's car reads │ 36 279 │.
   What will it say when she has gone these distances?

   (a) 20 miles   (b) 500 miles   (c) 2000 miles   (d) 40 000 miles

5 Write these lists of numbers in order, smallest first.

   (a) 6030   3060   3600   6300   6003

   (b) 27 418   24 781   28 741   24 817   27 841

   (c) 35 675   36 143   36 002   35 432   35 141

B Rounding
level 4

1 Round these numbers to the nearest ten.

   (a) 32   (b) 47   (c) 275   (d) 761   (e) 6074

2 Round these numbers to the nearest hundred.

   (a) 432   (b) 174   (c) 2195   (d) 58 790   (e) 4942

3 Round these numbers to the nearest thousand.

   (a) 2427   (b) 6840   (c) 24 591   (d) 87 258   (e) 20 695

4 The second highest mountain in the world is K2 at 8611 m.
   Round the number 8611 to the nearest

   (a) thousand   (b) hundred   (c) ten

5 Madagascar is one of the largest islands in the world.
   It has an area of 587 040 square kilometres.
   Round the number 587 040 to the nearest

   (a) hundred thousand   (b) ten thousand   (c) thousand

3 Shapes

You need a pair of compasses in section A, and squared paper in sections B, C, D and E.

A Circles

1 (a) Measure the radius of circle A.

(b) Measure the diameter of circle B.

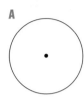 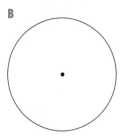

2 (a) Use a pair of compasses to draw a circle that has a radius of 5 cm.

(b) Label the centre of your circle.

(c) Label the circumference of your circle.

B Triangles

1 Choose one of these words to complete each sentence.

scalene   equilateral   isosceles

(a) A triangle that has edges all the same length is ....

(b) A triangle that has two edges the same length is ....

(c) A triangle with edges that are all different lengths is ....

2 Measure the edges of this triangle.
Which kind of triangle is it?

3 Draw a coordinate grid like this, going from 0 to 10 in both directions.

(a) (i) Plot the points given by the coordinates (2, 3), (5, 3) and (2, 6). Join them up with straight lines.

(ii) What kind of triangle have you drawn?

(b) (i) Plot the points (7, 2), (9, 7) and (6, 9) and join them up.

(ii) What kind of triangle have you drawn?

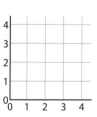

C Quadrilaterals

1 Look at these quadrilaterals.

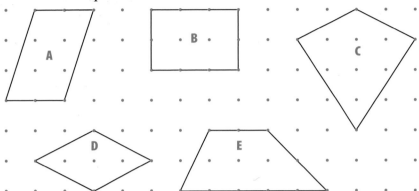

Give the letter for a quadrilateral that is

(a) a rhombus   (b) a trapezium   (c) a parallelogram   (d) a kite

2 Draw a coordinate grid that goes from 0 to 10 in both directions.

(a) (i) Plot the points $(1, 2)$, $(1, 5)$ and $(4, 5)$ on your grid.

(ii) Plot another point so that the four points are the corners of a square. Write down the coordinates of this point.

(b) (i) Plot the points $(3, 7)$, $(7, 7)$, $(9, 10)$ and $(5, 10)$ on your grid. Join the points up.

(ii) Write down the name of this shape.

3 Draw a rectangle.
Draw all the lines of symmetry on your rectangle.

4 (a) Copy and complete this diagram so that the dotted line is a line of symmetry.

(b) Write down the name of the complete shape.

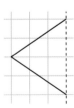

D More than four edges

1 (a) What is the name of this shape?

(b) How many sides does a pentagon have?

2 (a) From this set of shapes, list

    (i) all the hexagons

    (ii) all the octagons

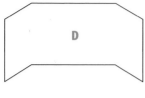 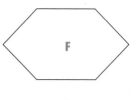

(b) Which of the shapes is

    (i) a regular pentagon

    (ii) a regular octagon

    (iii) a hexagon with only two lines of symmetry

3 For each diagram below

    (i) copy and complete the shape so that the dotted line is a line of symmetry

    (ii) write down the name of the complete shape

(a)    **(b)**   **(c)**

E Shading squares to give reflection symmetry

1 Copy this diagram.

Shade in four more squares so that the dotted line is the only line of symmetry.

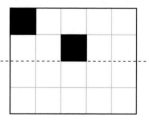

2 Copy this diagram.

    (a) Shade in three more squares so that the diagram has two lines of symmetry.

    (b) Draw the lines of symmetry on your diagram.

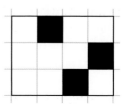

4 Adding and subtracting whole numbers

A Using written and mental methods level 3

1 Work these out.

   (a) $26 + 53$ (b) $47 + 18$ (c) $53 + 69$ (d) $235 + 451$

   (e) $762 + 45$ (f) $307 + 36$ (g) $68 - 37$ (h) $92 - 86$

   (i) $54 - 27$ (j) $159 - 26$ (k) $272 - 134$ (l) $463 - 175$

2 (a) There are 37 girls and 45 boys at a sports camp.
How many children are there at the sports camp altogether?

  (b) In the morning 32 of the children played tennis.
How many of the children did not play tennis?

  (c) In the afternoon 48 of the children took part in a softball tournament.
The rest went swimming.
How many children went swimming?

3 Class 9J are raising money for charity.

  (a) They make £17 from a cake stall and £58 from a sponsored silence.
How much money do they make altogether?

  (b) They give £50 to the local hospice and the rest to the school fund.
How much money do they give to the school fund?

4 Copy and complete these calculations, filling in the missing digits.

  (a)   ■ 1   **(b)**   5 ■   **(c)**   ■ 6   **(d)**   7 2
      + 4 ■   + ■ 8   − 2 ■   − 2 ■
      ——   ——   ——   ——
      7 7   9 0   3 2   4 4

5 (a) Find two numbers in the loop with a sum of

    (i) 100   (ii) 64   (iii) 62

  (b) Find two numbers in the loop with a difference of

    (i) 10   (ii) 17   (iii) 19

          47   26   9
      38   53
        57   21

6 Solve these number puzzles.

  (a)
I'm thinking of two numbers.
The difference between the numbers is 10.
The sum of the numbers is 50.
Find the two numbers.

  (b)
I'm thinking of two numbers.
The difference between the numbers is 6.
The sum of the numbers is 40.
Find the two numbers.

5 Listing

A Arrangements

1 Jamal has these three counters.
He wants to find all the different arrangements
of the three counters in a row.

Copy and complete this table to show all
the ways he can arrange the counters.

First	Second	Third
blue	white	yellow

2 Rachel has these three number cards.

(a) Copy and complete the table to show all
the different ways she can arrange the cards.

(b) What is the largest number she can make?

(c) How many numbers less than 450 can she make?

First	Second	Third
4	1	5

B Combined choices

1 Helen makes sandwiches with a choice of bread and fillings.

Bread
white
brown

Filling
egg
beef
cheese

(a) List all the possible combinations of bread and filling.
The list has been started for you.

(b) Samir doesn't like beef.
How many different types of sandwich
are there with no beef?

Bread	Filling
white	egg
white	beef

2 Maddie and Louise go to an arts club.
Each person can choose an activity from these three.

Dance		Acting		Photography

(a) List all the possible combinations of their choices.
The list has been started for you.

Maddie	Louise
dance	dance
dance	acting

(b) The photography activity is cancelled.
How many possible combinations of choices are there now?

6 Multiplying and dividing whole numbers

A Multiplying whole numbers
level 4

1 Work these out.

   (a) 17×3 (b) 19×5 (c) 24×8 (d) 36×9

   (e) 41×6 (f) 59×7 (g) 83×4 (h) 76×5

2 A carton holds 9 packets of cereal.
   How many packets will there be in 15 cartons?

3 A first class stamp costs 34p.
   How much does it cost for a book of 6 of these stamps?

4 It takes Sophie 21 minutes to walk one mile.
   At this rate, how long will it take her to walk 5 miles?

5 Mike swims 32 lengths every day.
   How many lengths does he swim in a week?

6 Tickets for a concert cost £22 each.
   How much does it cost for 9 of these tickets?

B Dividing whole numbers
level 4

1 Work these out.

   (a) $68 \div 2$ (b) $54 \div 3$ (c) $65 \div 5$ (d) $98 \div 7$

   (e) $99 \div 9$ (f) $92 \div 4$ (g) $78 \div 3$ (h) $96 \div 6$

2 Glass tumblers are sold in packs of 4.
   How many packs can be made from 64 tumblers?

3 Eggs are sold in boxes of 6.
   How many boxes can be filled from 90 eggs?

4 Work these out. Each answer has a remainder.

   (a) $71 \div 3$ (b) $83 \div 5$ (c) $86 \div 6$ (d) $97 \div 4$

5 Pens are sold in packs of 5.
   How many packs can be filled from 68 pens?

6 A tour guide at a museum can show 8 people around.
   How many tour guides are needed for a group of 90 people?

Mixed practice 1

1 Look at the numbers in this box.

2	10	16	22
26	30	38	41

   (a) Write down a number from the box that is
   - (i) odd
   - (ii) divisible by 3
   - (iii) a multiple of 4
   - (iv) a factor of 6

   (b) Find a pair of numbers from the box that add to give 42.

   (c) Find two numbers in the box with a difference of 14.

2 What is the name of a shape with eight straight sides?

3 (a) Use these cards to make the largest number you can.

   (b) What does figure 8 stand for in your number?

   (c) Write your number in words.

   (d) Round your number to the nearest hundred.

8 1 9 2

4 Work out 5613 − 200 in your head.

5 O is the centre of the circle.

   (a) Measure the length of the diameter of the circle.

   (b) (i) What kind of triangle is shown inside the circle?

   (ii) How many lines of symmetry does the triangle have?

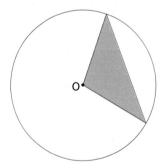

6 Work out 415 − 124.

7 A café offers a breakfast special.
You can choose a drink and a roll.

List all the possible combinations
of a drink and a roll.

Drinks
tea
coffee

Rolls
bacon
sausage
egg
ham

8 Small bottles of juice are sold in packs of 6.

   (a) How many bottles of juice are there altogether in 15 of these packs?

   (b) How many full packs can be made from 50 small bottles of juice?

9 Which is larger, 2008 or 2010?

10 Work out 84 ÷ 3.

7 Representing data

A Frequency charts and mode (for types of things)

1 This frequency chart shows the fish caught in a stretch of river during a Saturday.

   (a) How many roach were caught?

   (b) Which type was the mode?

   (c) How many fish were caught altogether?

On Sunday, a tally table was made of the types caught.

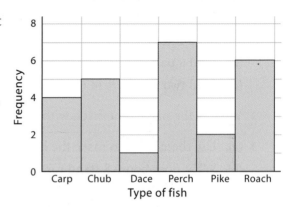

Type	Tally
carp	III
chub	JHT I
dace	I
perch	JHT III
pike	I
roach	JHT IIII

   (d) Which type was the mode on Sunday?

   (e) Draw a frequency chart for the Sunday data.

2 Prakesh carries out a survey of traffic crossing a bridge. He makes a list of the types of vehicle as they cross.
C = car, V = van, B = bus, L = lorry, M = motorbike

   (a) Make a tally table for the data.

   (b) Which type of vehicle was the mode?

   (c) Draw a frequency chart for the data.

V	V	L	C	C	M	C	C	C	L
C	V	L	V	M	C	B	C	V	C
L	V	L	C	M	M	C	C	M	C
B	V	C	V	C	C	L	L	L	M
V	C	V	V	L	C	C	C	L	V

B Dual bar charts
C Pictograms

1 This dual bar chart shows information obtained from a crime survey.

   It shows the percentage of crimes that were alcohol-related or drugs-related.

   (a) For which type of crime is the percentage that were alcohol-related the highest?

   (b) For which types of crime is the drugs-related percentage higher than the alcohol-related percentage?

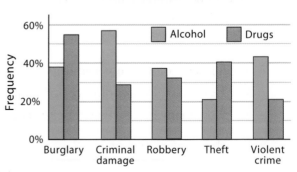

Source: Crime and disorder in Croydon audit 2001

2 This pictogram shows the number of different types of sandwich sold in a school canteen.

   (a) How many cheese sandwiches were sold?

   (b) How many vegetable sandwiches were sold?

   (c) Which filling was the mode?

   (d) How many sandwiches were sold altogether?

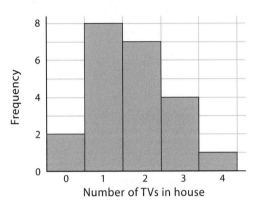

Sandwich filling   = 10 sandwiches

egg

cheese

meat

fish

vegetable

D Frequency charts and mode (for quantities)   level 4

1 Rajvi did a survey of the number of TVs in the houses in her street.

   Her results are shown in this frequency chart.

   (a) How many houses had 2 TVs?

   (b) Which number of TVs is the mode?

   (c) How many houses did Rajvi include in her survey?

2 Rajvi also recorded the number of people living in each house. Here are her results.

   3 4 2 4 3 1 5 3 4 5 3 6 3 4 3 2 1 6 4 2 2 5

   (a) Make a tally table for this data.

   (b) What is the mode?

   (c) Draw a frequency chart for the data.

3 A coach company runs 53-seater coaches from London to Brighton. The company keeps a record of the number of passengers in each coach.

   The table below shows the results for a week.

Number of passengers	48	49	50	51	52	53
Number of coaches	8	11	15	12	26	18

   (a) Which number of passengers is the mode?

   (b) How many coaches did the company run altogether in the week?

   (c) Draw a frequency chart for the data.

8 Fractions

1 Is a third of each of these squares shaded?
If not, say what fraction of the square is shaded.

(a) **(b)** **(c)** **(d)**

2 What fraction of each of these shapes is shaded?

(a) **(b)** **(c)** **(d)**

3 Make a copy of this rectangle on squared paper and shade $\frac{2}{5}$ of it.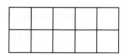

B Finding a fraction of a number

1 (a) Copy and complete this sentence.

To find $\frac{1}{3}$ of a number divide it by ...

(b) Work out

    (i) $\frac{1}{3}$ of 6   **(ii)** $\frac{1}{3}$ of 15   **(iii)** $\frac{1}{3}$ of 24   **(iv)** $\frac{1}{3}$ of 30   **(v)** $\frac{1}{3}$ of 75

2 Work these out.

    (a) $\frac{1}{2}$ of 48   **(b)** $\frac{1}{4}$ of 40   **(c)** $\frac{1}{5}$ of 35   **(d)** $\frac{1}{6}$ of 66   **(e)** $\frac{1}{8}$ of 72

3 Work these out.

    (a) $\frac{1}{5}$ of 50   **(b)** $\frac{2}{5}$ of 50   **(c)** $\frac{3}{5}$ of 50   **(d)** $\frac{4}{5}$ of 50

4 Work these out.

    (a) $\frac{2}{3}$ of 24   **(b)** $\frac{3}{4}$ of 28   **(c)** $\frac{5}{6}$ of 30   **(d)** $\frac{3}{8}$ of 160   **(e)** $\frac{3}{10}$ of 150

5 Find three pairs that give the same answer.

| $\frac{3}{4}$ of 80 | , | $\frac{4}{5}$ of 100 | $\frac{3}{4}$ of 120 | $\frac{2}{3}$ of 120 | $\frac{3}{5}$ of 150 | $\frac{2}{3}$ of 90 |

9 Decimal places

A One decimal place

1 What number does each arrow point to?

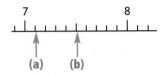 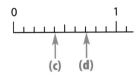 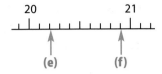

2 Which of these numbers are between 3 and 5?

3.5, 4, 2.6, 5.4, 3.9, 4.8, 6.1

3 Put each list of numbers in order, smallest first.

(a) 5.3, 4.8, 8.5, 6.3, 10.4   (b) 0.8, 1.3, 1, 0.3, 2

4 Diane was asked to draw a line 4.3 cm long.
This is her line.

(a) Measure the line. Is it longer or shorter than 4.3 cm?

(b) What is its length?

5 What is the number halfway between 7 and 8?

6 Here are the heights of five children.

    Pam 1.2 m   Charlie 1.3 m   David 0.8 m   Hayley 1.4 m   Harry 0.9 m

(a) Who is the tallest?   (b) Who is the shortest?

(c) Who is closest to 1 m tall?

B Two decimal places

1 What number does each arrow point to?

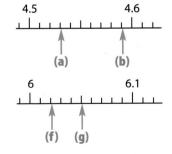 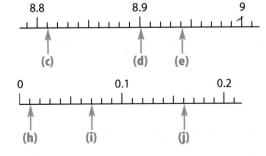

2 Decide whether each of these is true or false.

(a) 8.16 is larger than 8.24

(b) 3.85 is smaller than 3.9

(c) 2.8 is larger than 2.09

(d) 4.6 is smaller than 4.60

3 Put these cards in order from the smallest number to the largest.
What word do you get?

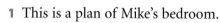

R	B	E	U	R	G
1.85	0.07	1.83	0.16	1.75	1.8

4 Which of these numbers are between 4.6 and 4.8?

3.7, 4.15, 4.72, 4.68, 4.07, 4.81, 4.7

5 Put these numbers in order, **largest** first.

6, 6.9, 5.6, 6.09, 5.49

6 In a guessing contest, some students guess the weight of a parcel.
The parcel weighs 1.78 kg.
Which of these guesses is the closest?

1.75 kg   1.8 kg   1.07 kg   1.83 kg   1.74 kg

7 What number is halfway between

(a) 10 and 11

(b) 1.8 and 1.9

(c) 7 and 7.1

(d) 4.9 and 5

c Decimal lengths

level 4

1 This is a plan of Mike's bedroom.

(a) What is the length of the bedroom?

(b) What is the width of the bedroom?

(c) What is the width of the doorway?

2 Mike buys a bed 2.15 m long.
Will the bed fit across the width of the room?

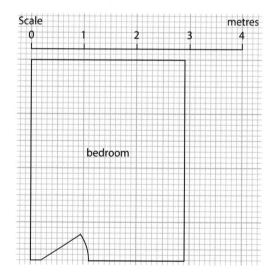

3 Mike's hall is 2.3 m wide.
Here are the widths of some carpets.

A 2.27 m   **B** 2.09 m   **C** 2.43 m   **D** 2.34 m   **E** 2.25 m

Which carpets are wider than the hall?

10 Median and range

A Median (for an odd number of data items)

1 A certain model of car is on sale at different car dealers.
The price at each dealer is shown here.

£8750   £9050   £8900   £9150   £9400   £9250   £9550   £9700   £9600

(a) Write the prices in order, lowest first.

(b) What is the median price?

2 Find the median of each of these sets of numbers.

(a) 203   187   160   230   206   163   227   150   177

(b) 145   162   157   151   162   150   148   148   153   159   147

B Median (for any number of data items)

1 Chris did a survey of students in part-time jobs.
As part of his survey he asked 12 students how much they were paid per hour.
Here are his results, in order.

£4.50   £4.70   £4.75   £4.80   £4.80   £5.10   £5.20   £5.20   £5.30   £5.45   £5.60   £6.00

Find the median rate of pay per hour.

2 Sam took an exam. It had ten questions, each marked out of 10.
Her marks were: 5   8   9   2   4   10   6   9   6   8

Find Sam's median mark for a question.

3 Find the median of each of these sets of numbers.

(a) 56   72   83   66   54   70   48   56

(b) 127   139   146   118   107   127   120   142   102

C Range

1 Eight bikes are for sale in a shop.
Their prices are: £122   £115   £108   £135   £124   £119   £130   £125

(a) Find the highest price.

(b) Find the lowest price.

(c) Find the range of the prices.

2 Find the median and range of each of these data sets.

(a) 96   88   112   127   94   86   117   120   103

(b) 212   236   194   203   175   201   187   242   219   177

D Comparing two sets of data

1 The table below shows the scores in the Rugby League matches
played during a period of two weeks.

Home team's score	23	42	20	10	20	18	28	20	30	18	10	25	32
Away team's score	24	6	22	4	8	2	47	30	22	35	22	22	6

(a) Find the median score for the home teams.

(b) Find the range of the scores of the home teams.

(c) Find the median score for the away teams.

(d) Find the range of of the scores of the away teams.

(e) Which scores, home or away, are higher on average?

(f) Which scores, home or away, are more spread out?

2 A club has two quiz teams, a men's team and a women's team.
Both teams took part in 12 competitions.
The scores were all out of 100.

Here are the men's and women's scores in the competitions.

Men	62	58	77	49	58	66	82	74	56	60	71	66
Women	68	64	69	55	72	70	72	69	53	58	69	74

(a) Find   (i) the men's median score   (ii) the range of the men's scores

(b) Find   (i) the women's median score   (ii) the range of the women's scores

(c) Which team, men or women, scored better on average?

(d) Which team's scores are more spread out?

3 The marks of the male and female students in an exam are arranged in order,
highest first.

Male students   93, 83, 79, 76, 72, 65, 61, 57, 55, 53, 50, 50, 47, 41

Female students   90, 88, 83, 80, 71, 69, 66, 65, 60, 57, 50, 46

(a) Find   (i) the males' median mark   (ii) the range of the males' marks

(b) Find   (i) the females' median mark   (ii) the range of the females' marks

(c) Which group, males or females, did better on average?

(d) Which group's marks are more spread out?

11 Mental methods 2

1 Write down the answers to these.
   (a) 42×10 (b) 257×100 (c) 10×190 (d) 96×1000 (e) 600×100
   (f) 2.8×10 (g) 3.1×100 (h) 5.29×10 (i) 0.35×1000 (j) 100×0.4

2 Write down the answers to these.
   (a) $500 \div 10$ (b) $4700 \div 100$ (c) $6200 \div 1000$ (d) $368 \div 10$ (e) $425 \div 100$
   (f) $96 \div 100$ (g) $170 \div 1000$ (h) $2.7 \div 10$ (i) $74 \div 100$ (j) $0.8 \div 10$

3 (a) A paper clip weighs 0.5 g.
      What is the weight of one hundred paper clips?

   (b) A drawing pin weighs 0.35 g.
      What is the weight of ten drawing pins?

4 (a) A pack of 10 notebooks costs £4.50.
      What is the cost of one notebook?

   (b) The pack of 10 notebooks is 8 cm thick.
      What is the thickness of one notebook?

5 A pen costs £1.30. What is the cost of 10 pens?

6 A box of 100 pencils costs £12. What is the cost of one pencil?

1 Work these out.
   (a) 3×20 (b) 2×400 (c) 4×60 (d) 6×800 (e) 9×700

2 Work these out.
   (a) 3×80 (b) 30×80 (c) 3×800 (d) 30×800 (e) 300×800
   (f) 300×50 (g) 40×30 (h) 3000×30 (i) 700×300 (j) 80×500

3 Work out the cost, in pounds, of
   (a) 100 crocus bulbs at 15p each   (b) 400 bluebell bulbs at 20p each
   (c) 40 lily bulbs at 80p each   (d) 500 hyacinth bulbs at 60p each

12 Solids, nets and views

A Solids and nets

level 4

1 Give the mathematical name of each solid below.
Choose a name from the loop.

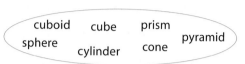

cuboid   cube   prism

sphere   cylinder   cone   pyramid

(a) 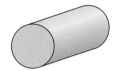   (b)   (c)

2 (a) Give the name of the solid made by each net below.

(i) 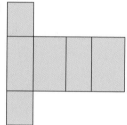   (ii) 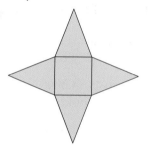   (iii)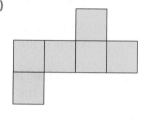

(b) For each solid, state the number of faces, vertices and edges.

B Views

1 Below are five views of this prism,
looking from the directions P, Q, R, S and T.
Match each view with its correct letter.

(a)   (b)   (c)

(d)   (e)

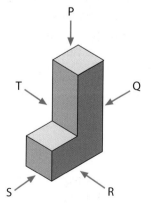

13 Weighing

A Grams and kilograms

1   Which completes each statement, grams or kilograms?

   (a) My suitcase weighs 15 _____ .   (b) My wallet weighs 300 _____ .

2   How many grams are there in each of these?

   (a) 4 kg   (b) 6 kg   (c) 13 kg   (d) $\frac{1}{4}$ kg   (e) $1\frac{1}{2}$ kg

3   Toby is baking a cake.
   He uses 250 g of sugar from a 1 kg bag.
   How much sugar is left in the bag?

4   Jean eats a 50 g portion of breakfast cereal every day.
   How many days will a 1 kg box of breakfast cereal last her?

5   What is the weight in kilograms of 5 packets of coffee that weigh 200 g each?

B Using decimals

1   How many grams are there in each of these?

   (a) 0.2 kg   (b) 2.5 kg   (c) 0.65 kg   (d) 1.25 kg   (e) 5.68 kg

2   (a) Copy and complete this table.
      It shows the weights of some birds
      in grams and in kilograms.

   (b) Which of the birds in the table
      is the heaviest?

   (c) Which is the lightest?

Bird	Weight in grams		Weight in kilograms
Herring gull	1200 g	=	
Common gull	450 g	=	
Mute swan		=	12 kg
Canada goose		=	4.5 kg

3   Here are the ingredients to make a large batch of shortbread.

   Copy the list but change all the amounts in grams to kilograms.

> **Shortbread**
> (makes 200 pieces)
> 3500 g butter
> 1500 g sugar
> 5000 g plain flour

4   Aaron has a bag of pasta that weighs 1.5 kg.
   He uses 600 g of it.
   How much pasta does he have left?

5   Put these weights in order, smallest first.   1.4 kg   4 kg   140 g   4 g   0.4 kg

Mixed practice 2

1. How many minutes are there in $\frac{3}{4}$ of an hour?

2. The heights (in cm) of ten one-year-old children are:

   69   73   65   78   70   68   72   74   67   74

   (a) Write this set of heights in order.

   (b) What is the median height of these children?

   (c) Find the range of these heights.

3. Work out $\frac{1}{5}$ of 60.

4. What number is halfway between 9 and 10?

5. This data shows the hours of sunshine in York each day during one July.

   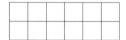

Hours of sunshine: July
6 4 5 6 7 3 4 6 5 2 6 1 5 3 7 5
4 6 8 5 3 5 4 2 1 3 5 6 8 2 4

   (a) Make a tally chart for this data.

   (b) On how many days were there 3 hours of sunshine?

   (c) Draw a frequency chart for the data.

   (d) What was the mode?

6. Copy this rectangle and shade $\frac{1}{3}$ of it to make a design with two lines of symmetry.

7. Lyra has 100 small bars of chocolate. Each bar weighs 35 g. What is the total weight of chocolate in kilograms?

8. Put these weights in order, smallest first.   250 g   0.75 kg $\frac{1}{2}$ kg   1.5 kg

9. Which is larger, 3.45 or 3.6?

10. Work these out.

    (a) 80×5 (b) 0.3×100 (c) 60×200 (d) $51 \div 100$

11. Penny has $\frac{1}{4}$ kg of butter.
    She uses 90 g.
    How much butter is left?

12. State whether this net gives a cuboid, a pyramid or a prism.

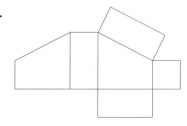

14 Time and travel

1 Write each of these times as

    (i) a 12-hour clock time using a.m. or p.m.   **(ii)** a 24-hour clock time

(a)    **(b)**   **(c)**

      Morning   Evening   Evening

2 Change these times into 24-hour clock times.

  (a) 10:35 p.m.   **(b)** 9:20 a.m.   **(c)** half past three in the afternoon

3 Change these times into 12-hour clock times using a.m. and p.m.

  (a) 07:00   **(b)** 11:55   **(c)** 17:30   **(d)** 21:40   **(e)** 00:10

1 (a) Copy and complete this diagram.

    (b) How long is it from 11:50 to 12:25?

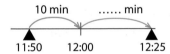

2 How long is it

  (a) from 09:45 to 10:05   **(b)** from 14:35 to 15:10   **(c)** from 6:30 p.m. to 7:25 p.m.

3 It takes Azam 25 minutes to drive to work.
   If he leaves the house at 8:15 a.m. what time does he arrive at work?

4 How long is it

  (a) from 07:30 to 10:15   **(b)** from 12:40 to 16:10   **(c)** from 9:50 a.m. to 12:15 p.m.

5 Karen goes to the cinema.

  (a) The film starts at 7:20 p.m. and finishes at 9:40 p.m.
     How long is the film?

  (b) It takes Karen 35 minutes to walk home afterwards.
     What time does she get home?

C Working out starting times

1 What time is it three hours before

   (a) 16:00

   (b) 9:30 a.m.

   (c) quarter past midday

2 (a) Copy and complete this diagram.

   (b) What time is it 35 minutes before 10:15?

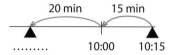

20 min   15 min

.... 10:00   10:15

3 What time is it

   (a) 15 minutes before 16:10

   (b) 40 minutes before 9:25 p.m.

4 It takes 25 minutes for Alex to walk to school.
   What is the latest he should leave home if school starts at 08:45?

5 Naomi is cooking a casserole which takes $1\frac{1}{2}$ hours to cook.
   She wants it to be ready at 6:15 p.m.
   What time should she put it in the oven?

D Timetables

1 Here is part of the timetable for trains
  on the North Yorkshire Moors Railway.

Pickering	11:20	12:20	14:20	15:20
Levisham	11:40	12:40	14:40	15:40
Newton Dale	11:49	12:49	14:49	15:49
Goathland	12:10	13:10	15:10	16:10
Grosmont	12:25	13:25	15:25	16:25

   (a) What time does the 12:20 train
       from Pickering arrive in Grosmont?

   (b) What time does the 14:40 train
       from Levisham arrive in Goathland?

   (c) How long do trains take to get from Pickering to Newton Dale?

   (d) Meena wants to arrive in Goathland by 2 p.m.
       What is the latest train she can get from Pickering?

2 Here is part of an evening bus timetable.

Christianfields	1632	1702	1732	1807
Kings Farm	1635	1705	1735	1810
Railway Station	1645	1715	1745	1820
Coldharbour	1658	1728	1758	1833
Pepper Hill	1703	1733	1803	1838

   (a) What time does the 1732 bus from
       Christianfields get to Pepper Hill?

   (b) What time does the 1705 bus from
       Kings Farm get to Coldharbour?

   (c) How long does the bus take to get
       from Christianfields to Coldharbour?

   (d) Leon arrives at the Railway Station at 6 p.m.

       (i) How long does he have to wait for the next bus to Pepper Hill?

       (ii) What time does he arrive at Pepper Hill?

15 Angle

You need an angle measurer for sections A and C.

A Drawing, measuring and sorting angles

1 Say how many degrees there are in
   (a) a quarter turn   (b) a half turn   (c) a right angle   (d) a three-quarter turn

2 Measure each of these angles

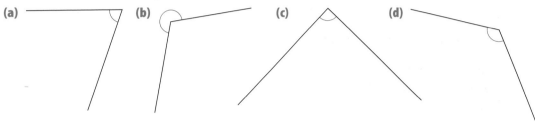

(a)   (b)   (c)   (d)

3 Say whether each angle in question 2 is acute, a right angle, obtuse or reflex.

4 Say whether each of these angles is acute, obtuse or reflex.
   (a) 45°   (b) 200°   (c) 95°   (d) 160°   (e) 85°

5 Draw each angle in question 4.

B Turning

In 12 hours, the hour hand of a clock makes a full turn, which is 360°.
So in one hour the hour hand turns through 360° ÷ 12, which is 30°.

1 Give the angle that the hour hand of a clock turns
   (a) between 12 noon and 2 p.m.   (b) between 6 p.m. and 10 p.m.
   (c) between 4 p.m. and 9 p.m.   (d) between 2 p.m. and 11 p.m.

2 What type of angle is each angle in question 1?

3 The hour hand of a clock points to 5.
   As time passes it turns 150° clockwise.

   (a) What number is it pointing to now?

   (b) How many hours have passed?

   (c) What clockwise angle must the hour hand now turn through
       to point to 5 once again?

C Angles in shapes

1 (a) Measure the angles of this triangle accurately.
The triangle is special in two ways.
What words describe the ways that it's special?

(b) What do the angles of this triangle add up to?

(c) Measure the triangle's sides accurately.
What do you notice about them?

2 (a) Measure the sides and angles of this shape carefully
and say which of them are equal.

(b) Give the special name of the shape.

D Estimating angles

In this section do not measure the angles.

1 This triangle is equilateral (it has all three sides equal).
Its angles are all equal to 60°.

Say whether each of the acute angles below
is less than 60° or more than 60°.

(a)   **(b)**   **(c)**   **(d)**

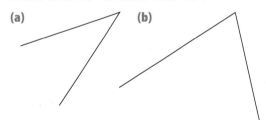
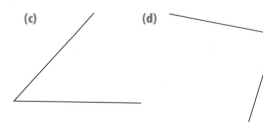

2 Write an estimate, to the nearest 10°, of each angle in question 1.

3 Which of the angles below is closest to 120° (twice 60°)?

A   B   C

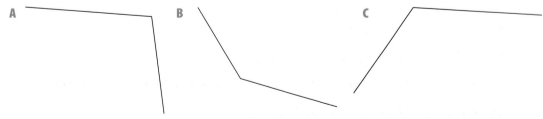

4 Write an estimate, to the nearest 10°, of each angle in question 3.

16 Length

A Centimetres and millimetres

1 Measure the lengths of these feathers.
   Write down their measurements in   **(i)** mm   **(ii)** cm

   (a)    **(b)**

2 **(a)** This table shows the lengths
   of some small birds.
   Copy the table and complete it.

   (b) Which of the birds is the shortest?

   (c) Which is the longest?

Bird	Length in millimetres		Length in centimetres
Goldcrest	90 mm	=	
Firecrest		=	8.5 cm
Blue tit	113 mm	=	
Long-tailed tit		=	14 cm

3 Put these lengths in order, shortest first.   3 cm   3 mm   1.3 cm   31 mm

B Metres and centimetres
C Kilometres

1 What completes each statement: cm, m or km?
   (a) My bookcase is 2.1 __ tall.   **(b)** My pencil is 14 __ long.
   (c) I walk 2 __ to work each day.   **(d)** My kitchen is 255 __ wide.

2 Write each of these lengths in centimetres.
   (a) 3 m   **(b)** $\frac{1}{2}$ m   **(c)** $1\frac{1}{4}$ m   **(d)** 2.6 m   **(e)** 0.85 m

3 Write each of these lengths in metres.
   (a) 3 km   **(b)** 10 km   **(c)** 4 km   **(d)** 1.5 km   **(e)** 0.75 km

4 Josie buys a piece of ribbon 1.5 m long.
   She cuts off a piece 90 cm long.
   How long is the piece that is left?

5 Put each set of lengths in order, shortest first.
   (a) 4 km, 400 cm, 40 mm, 0.4 m   **(b)** 102 cm, 12 m, 1.2 m, 12 cm

6 Doug is 1.6 m tall. Dora is 155 cm tall.
   Who is taller, and by how many centimetres?

17 Squares and square roots

A Square numbers and square roots
B Using shorthand

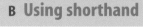

1 List all of the square numbers between 40 and 90.

2 Chris has 30 counters.
He makes the biggest square he can with them.
   (a) How many counters does he use?   (b) How many rows of counters are there?

3 Copy and complete these.
   (a) 4 squared is ■   (b) The square root of 81 is ■
   (c) ■^{2} = 9   (d) $\sqrt{■}$ = 7

4 Find the value of each of these.
   (a) 6^2 (b) $\sqrt{4}$ (c) 1^2 (d) $\sqrt{64}$ (e) 10^2

5 Use the clues to find the numbers.

   (a)
A square number
Less than 10
Greater than 5

   (b)
A square number
Less than 100
Ends in 5

   (c)
A square number
Less than 40
A multiple of 8

6 Which is bigger, 3^2 or $\sqrt{36}$?
Explain how you know.

C Using a calculator

1 Use your calculator to find the value of each of these.
   (a) 14 squared   (b) 23^2 (c) 18^2
   (d) the square root of 144 (e) $\sqrt{900}$ (f) $\sqrt{225}$

2 Use the clues to find the numbers.

   (a)
A square number
Between 300 and 400
An even number

   (b)
A square number
Between 700 and 800
Last digit 4

   (c)
A square number
Between 800 and 1000
A multiple of 3

3 These cards can be rearranged to make two square numbers.
What are the two square numbers?

5   2   6

18 Adding and subtracting decimals

A Adding decimals
level 4

1 Work these out in your head.

(a) 0.2 + 0.7   (b) 7.2 + 0.6   (c) £1.50 + £4   (d) £1.40 + £2.10

2 Tom buys a newspaper that costs £1.40 and a bottle of water that costs £0.75.
How much does he spend altogether?

3 Work these out.

(a) 3.5 + 5.8   (b) 9.6 + 2.7   (c) 15.9 + 3.2   (d) 42.6 + 17.5

4 Sara buys 0.35 kg of red grapes and 0.45 kg of green grapes.
What weight of grapes does she buy altogether?

5 Rita stands a bookcase 1.2 m wide next to a chest 0.85 m wide.
What is the total width of the bookcase and the chest?

6 Work these out.

(a) £6.28 + £3.50   (b) 5.16 + 10.75   (c) £17.65 + £11.47 (d) 21.85 + 8.3

B Subtracting decimals
level 4

1 Work these out in your head.

(a) 7.9 − 0.2   (b) 5.2 − 1   (c) £2 − £0.60   (d) £5.80 − £0.50

2 Martin has a piece of skirting board 2 m long. He cuts off 0.4 m.
What is the length of the remaining piece?

3 Sayed has a bag containing 3.5 kg of potatoes. He uses 1.2 kg of them.
What weight of potatoes does he have left?

4 Corinne has £8.85 in her purse. She buys a magazine that costs £3.90.
How much money does she have left?

5 Daniel spends £4.83 in the corner shop. He pays with a £10 note.
How much change does he get?

6 Work these out.

(a) 7.65 − 3.14   (b) 5.3 − 2.8   (c) 8.36 − 4.91   (d) 4.85 − 3.2

(e) 19.25 − 14.4   (f) 5.7 − 1.25   (g) 6.4 − 2.71   (h) 8 − 2.65

19 Mental methods 3

1 Work these out in your head.

(a) 18×4 (b) 26×5 (c) 34×4 (d) 5×51 (e) 4×35

(f) 5×48 (g) 42×4 (h) 63×5 (i) 4×125 (j) 82×5

2 Work these out in your head.

(a) $84 \div 4$ (b) $115 \div 5$ (c) $144 \div 4$ (d) $210 \div 5$ (e) $240 \div 4$

(f) $620 \div 5$ (g) $212 \div 4$ (h) $325 \div 5$ (i) $188 \div 4$ (j) $1250 \div 5$

3 Copy and complete this crossnumber puzzle.

ACROSS	DOWN
1 25×5	**1** $95 \div 5$
3 $248 \div 4$	**2** 4×14
6 $265 \div 5$	**4** $112 \div 4$
9 162×4	**5** 17×5
	7 $180 \div 5$
	8 22×4

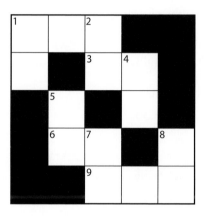

*4 (a) To multiply by 8 you can multiply by 2 and then by 2 again and then by 2 again.
Copy and complete this arrow diagram to work out 24×8.

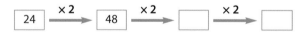

(b) Work these out.

(i) 13×8 (ii) 22×8 (iii) 17×8 (iv) 35×8

*5 (a) To divide by 8 you can divide by 2 and then by 2 again and then by 2 again.
Copy and complete this arrow diagram to work out $184 \div 8$.

```
  184   ÷2    92   ÷2   [   ]   ÷2   [   ]
```

(b) Work these out.

(i) $96 \div 8$ (ii) $128 \div 8$ (iii) $240 \div 8$ (iv) $448 \div 8$

Mixed practice 3

You need an angle measurer.

1. Fiona leaves for school at 7:50 a.m. and gets to school at 8:15 a.m.
   How long does it take her to get to school?

2. Write 1200 mm in cm.

3. Do these in your head.
   (a) 6.7 + 3   **(b)** 4 − 0.2   **(c)** 26 × 4   **(d)** 62 × 5   **(e)** 128 ÷ 4

4. Gina is going to see a film that starts at 19:40.
   She arrives at the cinema at a quarter to eight that evening.
   Is she on time for the film?

5. Write 2.3 metres in cm.

6.

   (a) Measure the length of the longest side of this shape.
   (b) **(i)** Which of the marked angles is obtuse?
         (ii) Estimate the size of this angle.
   (c) **(i)** Measure each angle.
         (ii) How many right angles are in this shape?
         (iii) What is the special name for this shape?

7. Sue needs to be at the station at 13:10.
   From her home, it takes her 35 minutes to get there.
   When does she need to leave her home?

8. **(a)** Is 30 a square number?   **(b)** Write down the square root of 25.

9. How many metres are there in $4\frac{1}{2}$ km?

10. Find the value of each of these.
    (a) 2 squared   **(b)** $\sqrt{9}$ **(c)** 4^2 **(d)** $\sqrt{100}$

11. Work out each of these.
    (a) 2.7 + 1.8   **(b)** 5.71 − 3.29   **(c)** 3.04 + 1.6   **(d)** 9.2 − 4.12

20 Number patterns

A Simple patterns
level 4

1 Here is a number pattern.

1   4   7   10   13   …

(a) What is the next number in the sequence?

(b) Explain how you worked out your answer.

2 Find the next two numbers in each number pattern.

(a) 25, 24, 23, 22, 21, …

(b) 11, 15, 19, 23, 27, …

(c) 40, 35, 30, 25, 20, …

(d) 1, 8, 15, 22, 29, …

3 Here is a number pattern.

4   8   12   16   20   …

(a) What is the next number in the sequence?

(b) Explain how you worked out your answer.

(c) Write down the first number in the sequence that is bigger than 40.

(d) Explain why 73 cannot be in this sequence.

B Further patterns

1 Here is a sequence.

5   10   20   40   80   …

(a) What is the next number in the sequence?

(b) Explain how you worked out your answer.

2 Find the next two numbers in each sequence.

(a) 5, 6, 8, 11, 15, …

(b) 128, 64, 32, 16, 8, …

3 (a) Copy and complete the first five lines of this number pattern.

$$5 \times 9 = 45$$
$$55 \times 9 = 495$$
$$555 \times 9 = \ldots\ldots$$
$$5555 \times 9 = 49\,995$$
$$\ldots\ldots \times 9 = \ldots\ldots\ldots$$

(b) Kyle thinks that $5\,555\,555 \times 9 = 499\,999\,995$ is a line from the same pattern. Explain why he is wrong.

21 Estimating and scales

A Estimating lengths

1 The diagram shows the plan of a tennis court.

   The tennis court is 10 m wide. Estimate its length.

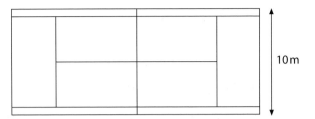

10 m

2 The height of the hippopotamus in these pictures is 1.5 m. Estimate the height of

   (a) the elephant

   (b) the hyena

   (c) the giraffe

B Reading and estimating from scales

1 For each of these scales

   (i)   say what one small division represents

   (ii) write down the number shown

(a)

(b)

(c)

(d)

2 These diagrams show some scales measuring in grams.
What weight does each show?

(a) 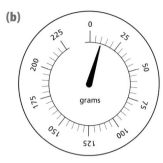 **(b)** 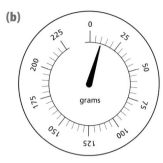 **(c)**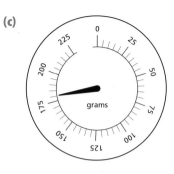

3 These scales have no small divisions marked.
Estimate the number each arrow points to.

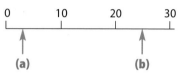 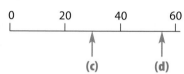

(a)　(b)　　　(c)　(d)

c Decimal scales

1 These diagrams show the scale on a spring balance
used in science lessons.

(a) What does each small division represent?

(b) What weight is shown in each case?

(i) 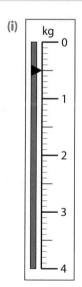 **(ii)**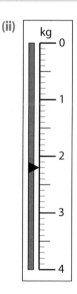

2 What number does each arrow point to?

(a)  **(b)**

(c)  **(d)**

22 Multiplying and dividing decimals

A Multiplying a decimal by a whole number

1 Work these out in your head.

   (a) 0.8×2 (b) 0.5×3 (c) 1.2×4 (d) 3.2×3 (e) 1.1×5

2 Work these out.

   (a) 1.9×5 (b) 8.3×4 (c) 7.8×3 (d) 9.3×6 (e) 5.2×7

3 Choose numbers from the loop to make these multiplications correct.

   (a) $0.9 \times \blacksquare = 2.7$ (b) $8 \times \blacksquare = 2.4$

   (c) $0.5 \times \blacksquare = 4.5$ (d) $\blacksquare \times 2 = 3.2$

   (e) $\blacksquare \times 0.4 = 2.4$ (f) $4 \times \blacksquare = 2.4$

   9   6   0.3
   1.6   0.6
      0.9   3

4 A carton contains 0.3 litres of orange juice.
   How much orange juice is there in 6 of these cartons?

5 A bag of chicken drumsticks weighs 1.6 kg.
   What is the weight of 8 of these bags?

B Dividing a decimal by a whole number

1 Work these out in your head.

   (a) $4.2 \div 2$ (b) $3.6 \div 3$ (c) $8.4 \div 4$ (d) $15.5 \div 5$ (e) $14.8 \div 2$

2 Work these out.

   (a) $7.2 \div 3$ (b) $9.5 \div 5$ (c) $11.6 \div 4$ (d) $13.5 \div 3$ (e) $16.2 \div 6$

3 A piece of cheese weighing 1.5 kg is cut into three equal pieces.
   How much does each piece weigh?

4 Find three matching pairs of divisions that give the same answer.
   Which is the odd one out?

   A $38.4 \div 6$ **B** $31.5 \div 5$ **C** $27.6 \div 3$ **D** $25.6 \div 4$ **E** $36.5 \div 5$ **F** $36.8 \div 4$ **G** $21.9 \div 3$

5 Work these out. Write each answer as a decimal.

   (a) $11 \div 2$ (b) $18 \div 4$ (c) $23 \div 5$ (d) $54 \div 4$ (e) $68 \div 8$

23 Area and perimeter

You need tracing paper in section A.

A Shapes on a grid of centimetre squares
level 4

1 These shapes are on a grid of centimetre squares.
For each shape find **(i)** its area and **(ii)** its perimeter, using the correct units.

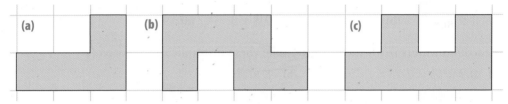

2 Find the area of each shape.

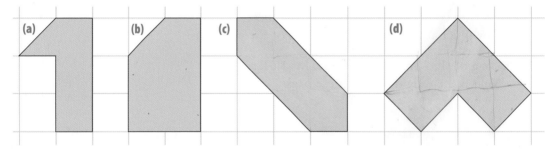

3 Estimate the area of this shape.
You can cover it with tracing paper to mark
the squares you count (so you don't have to
mark this booklet).

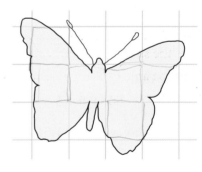

B Area of rectangle and right-angled triangle

1 Find the area of each of these rectangles.

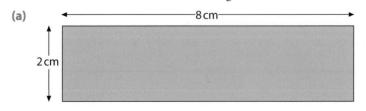

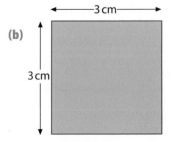

2 A certain rectangle has an area of 48 cm².
Its sides are whole numbers of centimetres long.
What lengths could its sides be? Give as many possibilities as you can.

3 These rectangles are not drawn full size. Find the area of each one.

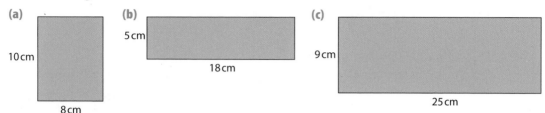

(a) 10 cm, 8 cm

(b) 5 cm, 18 cm

(c) 9 cm, 25 cm

4 Find the area of each blue triangle.

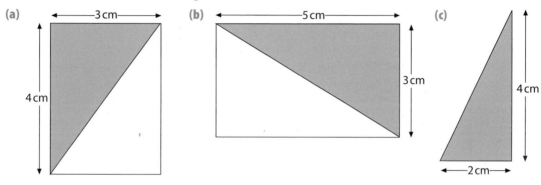

(a) 3 cm, 4 cm

(b) 5 cm, 3 cm

(c) 4 cm, 2 cm

c Area of a shape made from simpler shapes

1 (a) Do a sketch of this L-shape.
Split the shape into two rectangles.

(b) Find the area of the whole shape.

(c) Find the perimeter of the whole shape.

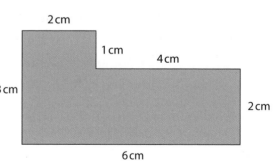

2 cm, 1 cm, 4 cm, 3 cm, 2 cm, 6 cm

2 This diagram shows the plan of a bedroom.
Find the area and perimeter of the bedroom.
Use the correct units in your answers.

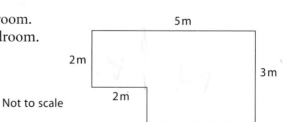

5 m, 2 m, 3 m, 2 m

Not to scale

D Using decimals

1

The plan above shows the upstairs of a house.

- (a) Without using a calculator, find the area of each of these.
  - (i) Bedroom 1   (ii) Bedroom 2
- (b) Use a calculator to find the area of each of these.
  - (i) The bathroom   (ii) The toilet   (iii) The box room

2 Find the area of each of these right-angled triangles.
They are not drawn to scale.

(a)

(b)

24 Probability

A The probability scale

1 Ruby has these cards.
She picks a card at random.

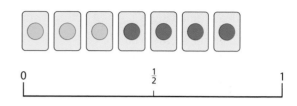

Copy this probability scale and mark an arrow to show, roughly,

(a) the probability that Ruby picks a yellow card

(b) the probability that she picks a blue card

2 Which arrow on this probability scale goes with each statement below?

(a) It rarely snows on Christmas Day.

(b) A dice is equally likely to show an even number as an odd number.

(c) The sun is certain to rise tomorrow.

(d) Rovers are very likely to win tomorrow.

B Equally likely outcomes

1 A bag contains these counters.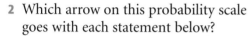

Carla picks a counter at random from the bag.

(a) (i) Which colour is she most likely to pick? (ii) Which is she least likely to pick?

(b) What is the probability that she picks

(i) a yellow counter (ii) a blue counter (iii) a white counter

(c) Bruno plays a game with the bag of eight counters.
He takes a counter at random.
If it is either white or yellow, he wins a prize.

What is the probability that he wins a prize?

2 A bag contains these shapes.

Gemma picks a shape at random from the bag.

What is the probability that she picks

(a) a triangle (b) a square (c) a circle (d) a grey shape

(e) a blue shape (f) a white shape (g) either a blue or a grey shape

3 Farnaz picks a shape at random from this set.

What is the probability that she picks

(a) a triangle (b) a square (c) a blue shape (d) a yellow shape

(e) a blue triangle (f) a yellow square (g) a blue square (h) a yellow triangle

(i) either a blue triangle or a yellow square

4 Danny picks a card at random from this set.

What is the probability that he picks

(a) the number 5 (b) an even number (c) an odd number

(d) a number less than 7 (e) a number greater than 7 (f) the number 10

5 The arrow on this spinner is spun.

Find the probability that the arrow stops on

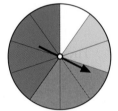

(a) white (b) yellow

(c) grey (d) blue

(e) yellow or grey (f) grey or blue

(g) white, grey or blue (h) yellow, grey or blue

6 Pam picks a card at random from this set.

What is the probability that she picks a card with

(a) three yellow spots (b) only one yellow spot

(c) exactly two yellow spots (d) a mixture of yellow and blue spots

(e) either one or two yellow spots (f) either two or three yellow spots

*7 A book has 15 pages, numbered from 1 to 15.
Rob opens the book at a random page.

What is the probability that the page number is

(a) odd (b) even

Mixed practice 4

1 Here is a number pattern.

   3   10   17   24   31   38   …

   (a) What is the next number in the pattern?

   (b) Explain how you worked out your answer.

2 What weight does each arrow point to?

   (a)

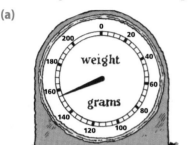

   (b)

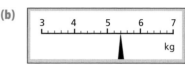

3 A piece of wood is 1.4 m long.
   What is the total length of 5 of these pieces of wood?

4 Leo puts these cakes in a box.
   He then chooses one at random.
   What is the probability that the cake

   (a) has pink icing   **(b)** has a cherry

   (c) has white icing but no cherry

5 Find the next two numbers in the sequence: 1, 3, 6, 10, 15, …

6 What is the probability of rolling an even number when
   an ordinary dice is rolled once?

7 Work out 20.1 ÷ 3.

8 This diagram shows a garden.

   (a) Find the perimeter of the pond.

   (b) Find the area of

       (i) the patio

       (ii) the pond

       (iii) the lawn

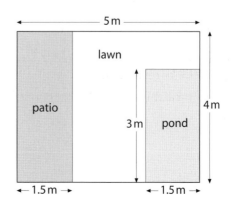

25 Enlargement

You need centimetre squared paper for section A.

A Enlargement on squared paper

1 On centimetre squared paper, make a three times enlargement of each shape.

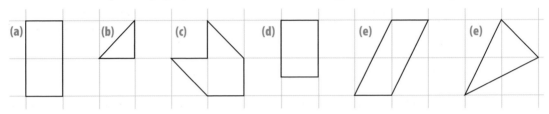

B Scale factor

1 Shapes Q and R are enlargements of shape P.

(a) Measure lengths of sides to find the scale factor of the enlargement from shape P to shape Q.

(b) What is the scale factor from shape P to shape R?

(c) If shape P is enlarged by scale factor 4, what will be the perimeter of the new shape?

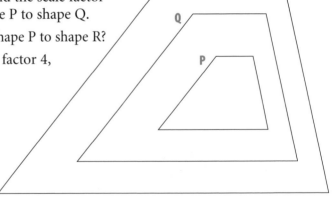

2 This is an enlargement of a smaller triangle. Side AB was 6.5 cm long in the smaller triangle.

(a) What scale factor has been used for the enlargement?

(b) What were the lengths of AC and BC in the smaller triangle?

26 Negative numbers

A Putting temperatures in order

1 (a) Write down the temperature, in °C, shown on each of these thermometers.

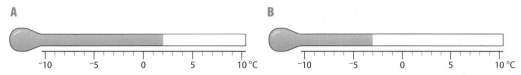

(b) Which thermometer shows the colder temperature?

2 This table shows the temperatures in some European cities one day in January.

City	Temperature
Bratislava	⁻3 °C
Hamburg	⁻2 °C
Milan	1 °C
Paris	0 °C
Stockholm	⁻4 °C
Zurich	⁻6 °C

(a) Which city was coldest?

(b) Which city was warmest?

(c) How many cities had temperatures below 0 °C?

(d) Write the temperatures in order, starting with the lowest.

3 Write these lists of temperatures in order, starting with the lowest.

(a) 3 °C, 0 °C, ⁻8 °C, ⁻3 °C, 8 °C   **(b)** 10 °C, ⁻20 °C, ⁻5 °C, 0 °C, ⁻15 °C

B Temperature changes

1 At 6 a.m. the temperature was ⁻3 °C.
At 10 a.m. the temperature was 2 °C.
How many degrees warmer was it at 10 a.m. than at 6 a.m.?

2 On Friday the temperature at noon was ⁻1 °C.
On Saturday the temperature at noon was 4 degrees warmer.
What was the temperature at noon on Saturday?

3 (a) What temperature does this thermometer show?

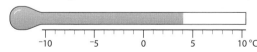

(b) The temperature falls by 7 degrees.
What temperature is it now?

4 Fergus recorded the temperature at different times one day in January.

Time	6 a.m.	9 a.m.	noon	3 p.m.	6 p.m.	9 p.m.
Temperature	⁻4 °C	⁻1 °C	2 °C	3 °C	1 °C	⁻2 °C

(a) At what time was the temperature highest?

(b) What was the lowest temperature?

(c) By how many degrees did the temperature rise between 6 a.m. and 9 a.m.?

(d) By how many degrees did the temperature fall between 6 p.m. and 9 p.m.?

(e) Copy and complete this sentence.

At 6 p.m. the temperature was …… degrees ………… than at 6 a.m.

(f) At midnight it was 3 degrees colder than it was at 9 p.m.
What was the temperature at midnight?

5 The highest temperature ever recorded in the UK is 39 °C.
The lowest temperature ever recorded in the UK is ⁻27 °C.
What is the difference between these temperatures?

c Negative coordinates

Use the grid to answer the questions below.

1 (a) What letter is at the point

    (i) (3, 1)   **(ii)** (2, ⁻2)

  (b) Write down the coordinates of

    (i) D   **(ii)** P

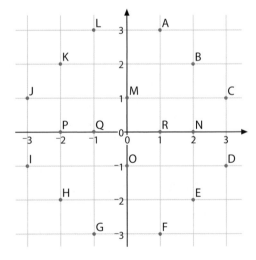

2 Write down the letters given by each set of coordinates below.
Rearrange the letters to give a colour.

  (a) (⁻2, 2)   (1, 3)   (3, 1)   (2, 2)   (⁻1, 3)

  (b) (2, ⁻2)   (⁻1, ⁻3)   (2, 0)   (2, ⁻2)   (1, 0)

3 Write down the coordinate codes for each of these.

  (a) PINK   **(b)** ORANGE

27 Mean

A Finding the mean of a data set

1 Patrick has five plants. Their heights are 76 cm, 81 cm, 66 cm, 68 cm, 69 cm.

   (a) Find the total height of the plants.   (b) Find the mean height.

2 Neeta is doing a survey of TV advertising.
   She times the adverts that appear between parts of a programme.
   Their lengths in seconds are   25   18   15   28   22   26   20

   Find the mean length of these adverts.

3 Find the mean of each of these data sets.

   (a) 6 7 3 6 3 3 7   (b) 34 28 22 39 57 50 37 37

   (c) 7.7 8.2 8.0 6.6 7.5   (d) 1.8 0.7 2.0 1.6 0.9 1.3 0.6 1.4 2.3

4 The ages of the children in a choir are 7, 8, 8, 9, 9, 11, 11, 13, 14.

   (a) Find the mean age of the choir.

   (b) The two oldest members leave. Find the new mean age of the choir.

B Comparing two sets of data

1 Dilip has two tomato plants, A and B, in different parts of his garden.
   When the tomatoes are fully grown, he weighs them.
   Here are the weights in grams.

   Plant A   27 29 33 35 28 28 33 31 28 38
   Plant B   30 27 24 30 32 22 25 26 36

   (a) Find the mean weight of the tomatoes from   (i) plant A   (ii) plant B

   (b) Which plant had heavier tomatoes, on average?

   (c) Find the range of the weights of the tomatoes from   (i) plant A   (ii) plant B

   (d) For which plant were the weights more spread out?

2 The prices of houses for sale in two streets are as follows. (£135k means £135 000.)

   Oak Avenue   £135k   £150k   £129k   £155k   £110k   £160k   £117k   £140k
   Duck Hill   £123k   £152k   £144k   £127k   £129k   £124k   £132k

   (a) In which street are the prices higher, on average?
       Show how you get your answer.

   (b) In which street are the prices more spread out?
       Show how you get your answer.

28 Starting equations

1 Find the missing number in each arrow diagram.

(a) 14 $\xrightarrow{\div 2}$?

(b) 15 $\xrightarrow{-6}$?

(c) ? $\xrightarrow{+1}$ 8

(d) ? $\xrightarrow{\div 3}$ 5

(e) 10 $\xrightarrow{-\,?}$ 6

(f) 4 $\xrightarrow{\times\,?}$ 20

2 Solve each of these.

(a)
> I think of a number.
> I double it.
> The result is 16.
> What number did I think of?

(b)
> I think of a number.
> I subtract 30.
> The result is 6.
> What number did I think of?

(c)
> I think of a number.
> I divide by 3.
> The result is 7.
> What number did I think of?

1 Find the missing number in each of these.

(a) $\heartsuit + 4 = 7$

(b) $\bigstar - 3 = 6$

(c) $\ast \times 5 = 30$

(d) $12 - \maltese = 10$

(e) $\vdots \div 4 = 2$

(f) $8 \times \blacklozenge = 24$

(g) $15 \div \ast = 5$

(h) $\circledast - 10 = 15$

(i) $\star + 8 = 20$

2 Find the missing number in each of these.

(a) $\blacksquare + \blacksquare = 10$

(b) $\star + \star + \star = 12$

(c) $\blacktriangle + \blacktriangle + \blacktriangle = 60$

3 Find the number that each letter represents.

(a) $n + 2 = 6$

(b) $x - 1 = 8$

(c) $p \div 2 = 4$

(d) $y \times 4 = 16$

(e) $w - 10 = 7$

(f) $6 \times m = 24$

(g) $5 + a = 9$

(h) $50 \div q = 5$

4 Solve each of these equations.
Write each solution as '$n = \ldots$'.

(a) $n + 10 = 25$

(b) $n - 6 = 12$

(c) $3n = 27$

(d) $n \div 2 = 10$

(e) $40 = 10n$

(f) $10 - n = 3$

(g) $16 = 12 + n$

(h) $30 \div n = 6$

(i) $14 + n = 20$

(j) $25 = 30 - n$

(k) $2n = 28$

(l) $3 = 12 \div n$

29 Finding your way

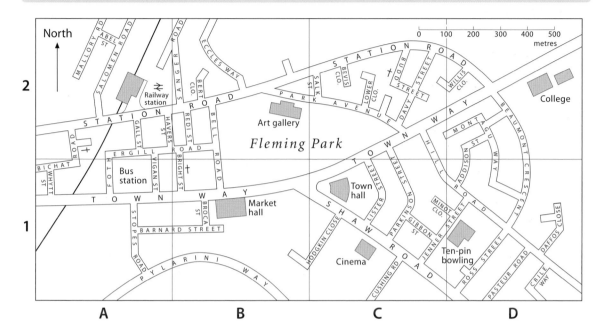

1 Which public building is in square B2?

2 In which square is **(a)** the town hall   **(b)** the college   **(c)** the market hall

3 When you walk south-west along Ross Street (D1) to the end of the street
   do you turn left or right to go towards the cinema?

4 Louise stands in the middle of Fleming Park. She turns slowly and sees
   the art gallery, the market hall and the town hall, in that order.
   Is she turning clockwise or anticlockwise?

5 From the railway station turn left into Station Road.
   When you get to Park Avenue turn right into it.
   At the end turn left into Town Way then immediately right into Hill Road.
   Take the first turn on the right and the place you want is on the left.

   (a)   Which place have these instructions taken you to?

   (b)   Use the scale and the edge of a piece of paper to estimate the length of this journey.

6 Give directions for someone who wants to travel

   (a)   from the cinema to the bus station

   (b)   from the town hall to the church in Budd Street (C2)

Mixed practice 5

1

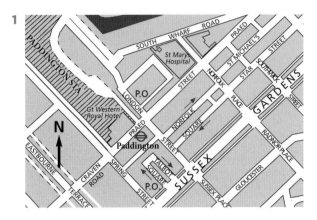

(a) You come out of St Mary's hospital and turn left into Praed Street. Which compass direction are you facing?

(b) From Talbot Square you go north-east along Sussex Gardens. You take the second left and first right. Where are you?

(c) Give directions to get from Paddington Tube station (marked ⊖) to Radnor Place.

2 Shape B is an enlargement of shape A. What is the scale factor of the enlargement?

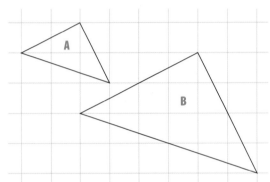

3 The table shows the temperature on January 1st at noon in five cities.

London	Moscow	Sydney	Oslo	Mumbai
1 °C	⁻8 °C	12 °C	⁻5 °C	20 °C

(a) Which city had the highest temperature?

(b) Which city had the lowest temperature?

(c) Sydney was hotter than Oslo. By how many degrees?

(d) By 6 p.m., the temperature in Moscow had dropped 5 degrees. What was the temperature in Moscow at 6 p.m.?

4 Solve these equations.

(a) $n + 3 = 10$ (b) $n - 1 = 5$ (c) $3n = 15$ (d) $n \div 2 = 8$

5 A group of students collected money for charity. The amounts they collected were

   £10   £12   £8   £9   £6   £15   £17   £11

(a) Find the mean amount collected.

(b) Find the range of the amounts.

30 Volume

A Counting cubes

1 These shapes are made out of centimetre cubes.
Find the volume of each shape, giving the correct units in your answer.

(a) 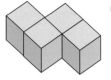 (b) 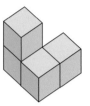 (c) 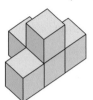 (d)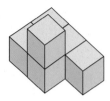

2 This prism is made out of centimetre cubes.
Find the volume of

  (a)   the orange part

  (b)   the blue part

  (c)   the whole prism

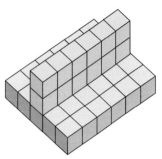

3 Some centimetre cubes have been piled up in layers.
Each layer is a rectangle.

  (a)   What is the name for the complete shape?

  (b)   How many cubes are there in the top layer?

  (c)   How many layers are there?

  (d)   What is the volume of the whole shape?

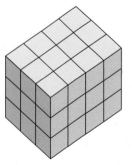

4 These cuboids are made out of centimetre cubes.
Find the volume of each one.

(a) 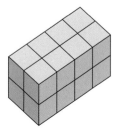 (b) 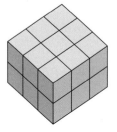 (c)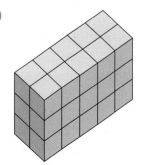

31 Evaluating expressions

A Simple substitution

1 What is the value of each expression when $n = 5$?

(a) $n + 1$ (b) $n + 6$ (c) $n - 2$ (d) $n - 5$

2 What is the value of each expression when $x = 6$?

(a) $2x$ (b) $4x$ (c) $x \div 2$ (d) $x \div 3$

3 Evaluate each expression when $n = 8$.

(a) $n + 3$ (b) $n - 4$ (c) $3n$ (d) $n \div 4$

B Rules for calculation

1 Work out each of these.

(a) $(6 + 4) \times 2$ (b) $6 + 4 \times 2$ (c) $5 \times (2 + 3)$ (d) $5 \times 2 + 3$

(e) $(6 + 4) \div 2$ (f) $6 + 4 \div 2$ (g) $14 \div 7 + 7$ (h) $2 + 10 \div 5$

2 Work out each of these.

(a) $(7 - 3) \times 2$ (b) $7 - 3 \times 2$ (c) $4 \times (5 - 3)$ (d) $4 \times 5 - 3$

(e) $(9 - 6) \div 3$ (f) $9 - 6 \div 3$ (g) $10 \div 2 - 3$ (h) $20 - 10 \div 2$

3 Work out each of these.

(a) $(2 + 5) \times (10 - 7)$ (b) $3 \times (1 + 5) - 10$ (c) $(13 - 3) \div (4 + 1)$

4 Work out each of these.

(a) $\dfrac{15}{3} + 2$ (b) $10 + \dfrac{6}{2}$ (c) $\dfrac{12 + 9}{3}$ (d) $\dfrac{10 - 4}{2}$

(e) $\dfrac{16}{4} - 3$ (f) $8 - \dfrac{9}{3}$ (g) $\dfrac{8 + 4}{1 + 5}$ (h) $\dfrac{17 - 3}{2 + 5}$

C Substituting into linear expressions

1 What is the value of each expression when $p = 4$?

(a) $6p + 1$ (b) $6(p + 1)$ (c) $2p - 3$ (d) $2(p - 3)$

2 Evaluate each expression when $n = 12$.

(a) $\dfrac{n}{3} + 6$ (b) $\dfrac{n + 6}{3}$ (c) $\dfrac{n - 4}{2}$ (d) $\dfrac{n}{2} - 4$

3 (a) Copy and complete this table to show the value of each expression for some different values of n.

	$n = 2$	$n = 4$	$n = 6$	$n = 8$
$2n + 12$				
$3n - 2$			16	
$3(n + 2)$				
$2(n - 1)$	2			
$\dfrac{n + 10}{2}$				
$\dfrac{n}{2} - 1$				

(b) (i) Which expression has a value of 6 when $n = 4$?

(ii) Which expression has the smallest value when $n = 8$?

(iii) Which expressions have the same value when $n = 6$?

D Formulas in words

1 Sue uses this rule to estimate how long a walk will take her in hours.

$$\text{time in hours} = \frac{\text{number of miles}}{2}$$

Roughly how long will Sue take to walk

(a) 4 miles   (b) 8 miles   (c) 10 miles

2 Pat's Pizzas use this rule to work out the cost of their pizzas in pence.

$$\text{cost in pence} = 25 \times \text{number of toppings} + 350$$

Work out the cost of a pizza with

(a) 1 topping   (b) 4 toppings   (c) 7 toppings

3 Jules uses this rule to estimate how long a walk will take him in minutes.

$$\text{time in minutes} = 12 \times \text{number of kilometres} + 30$$

Roughly how long will Jules take to walk

(a) 5 kilometres   (b) 8 kilometres   (c) 10 kilometres

4 Youth Activities use this rule to find the number of instructors they need for a group of students.

$$\text{number of instructors} = \frac{\text{number of students}}{4} + 1$$

How many instructors do they need for 20 students?

5 Ms North is a primary school teacher who gives gold and silver stars for effort.
She uses this rule to work out points for effort.

number of points = 5 × number of gold stars + number of silver stars

Hardeep gets 10 gold stars and 24 silver stars for effort.
How many points is this?

E Formulas without words

1 At Cycle Hire they charge £7 an hour to hire a bike.
You can use this formula to work out how much to pay.

$C = 7H$

C is the total cost in pounds and H is the number of hours.

What is the total cost if you hire a bike for

(a) 3 hours   (b) 6 hours   (c) 1 hour

2 A fabric shop uses this rule to work out the length of fabric needed to cover a sofa.

$L = 4S + 60$

L is the length of fabric in centimetres and S is the length of the sofa in centimetres.

What length of fabric do you need if the length of the sofa is

(a) 120 cm   (b) 160 cm   (c) 170 cm

3 Sam uses this formula to work out the height of a roof.

$H = \dfrac{R}{2} + 25$

H is the height of the roof in centimetres and
R is the length of the rafter in centimetres.

Calculate the height of the roof if the rafter is

(a) 350 cm   (b) 480 cm   (c) 550 cm

rafter

height

4 Alex uses this rule to work out the total height of a stack of chairs.

$H = 11N + 84$

H is the total height in centimetres and
N is the number of chairs in the stack.

What is the total height of

(a) 2 chairs   (b) 7 chairs

32 Estimating and calculating with money

A **Solving problems without a calculator**   level 4
B **Estimating answers**   level 4

1 (a) Work out the total cost for each of these groups to use
the swimming pool.
Show clearly how you get your answer each time.

Swimming pool	
Adult swim	£3.50
Child swim	£2.00
Senior swim	£2.50

   (i)   Three children   (ii) Two adults and one child

   (iii) Three adults   (iv) One adult and one senior

   (b) Jermaine buys two adult tickets. He pays with a £10 note.
   How much change does he get?

   (c) The swimming pool has a special family ticket for £9.00.
   David and Kate take their **two** children swimming.

   How much cheaper is it to buy a family ticket instead
   of separate tickets for two adults and two children?

   SPECIAL OFFER
   Family swim £9.00
   (up to 2 adults + 3 children)

2 It costs £2.95 to use a soft play area.

   (a) Roughly how much does it cost for 3 children to use the soft play area?

   (b) Michelle brings 5 children to use the soft play area.
   Does it cost more or less than £15 for 5 tickets?
   Explain how you know.

C **Using a calculator**   level 4

1 Work out the cost of

   (a) 7 bags of sweets at £1.15 each

   (b) 12 boxes of chocolates at £3.50 for each box

   (c) A box of mints for £2.79 and a bag of toffees for £1.05

2 Mini Easter eggs are sold in packs of 15.
Milo needs 50 mini eggs.

   (a) How many packs does he need to buy?

   (b) A pack of mini eggs costs £1.45.
   How much will Milo spend on the mini eggs?

3 Lina buys 5 bars of chocolate.
How much money does she save if she buys the multipack
instead of buying the bars separately?

   SAVE!
   Chocolate bars £1.29 each
   Multipack of 5 only £5.50

33 Capacity

A Litres, millilitres and other metric units

1 A jug of water sits on a table in a café.
   Which of these could be the capacity of this jug?

   1.5 ml   15 ml   1.5 litres   15 litres

2 A container holds 3000 ml of milk.
   How many litres of milk is this?

3 Kate has a 2 litre bottle of cola.
   How many 200 ml glasses can she fill?

4 Jason uses 600 ml of milk from a container that holds 1.5 litres.
   How much milk is left?

5 How many millilitres are there in $\frac{1}{4}$ of a litre?

6 Write 2500 ml in litres.

7 How many millilitres are there in each of these?

   (a) 6 litres   (b) 15 litres   (c) 4.5 litres   (d) 0.6 litres

8 How many litres are there in each of these?

   (a) 4000 ml   (b) 8000 ml   (c) 3500 ml   (d) 1750 ml

9 Put these volumes in order, smallest first.

   $\frac{1}{2}$ litre   250 ml   1.5 litres   70 ml   1700 ml

10 Which metric unit completes each statement?

   (a) The distance round a duckpond is 200 ……… .

   (b) Jason weighs 85 ……… .

   (c) Eve is 140 ……… tall.

   (d) My largest saucepan holds 5 ……… of liquid.

   (e) Kayley walks a total distance of 6 ……… to and from school each week.

34 Drawing and using graphs

A Tables and graphs

1 Amanda is collecting liquid in a chemistry experiment.

When she starts timing she has 2 ml of liquid in the cylinder.
Each minute she collects 3 more millilitres of liquid.

(a) How much liquid will she have after 2 minutes?

(b) Copy and complete this table.

Time in minutes	0	1	2	3	4	5
Liquid collected in ml	2	5				

(c) On graph paper draw and label axes like these.
Then plot the points from your table.

(d) Join the points you have plotted.
Extend the line they make.

(e) Use your graph to say how much liquid
Amanda will have after 7 minutes.

(f) How many minutes will it take until she
has 20 ml of liquid?

(g) How much liquid will she have after $3\frac{1}{2}$ minutes?

(h) The cylinder she is collecting the liquid in
can only hold 25 ml.
How long will it be until the cylinder is full?

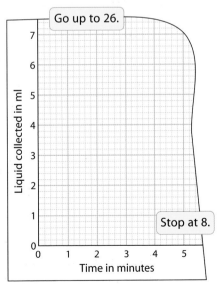

2 A small pond must be emptied before cleaning.

The pond contains 2500 litres of water.
180 litres of water are pumped out of the pond every minute.

(a) How much water is left in the pond after 1 minute?

(b) How much water is left in the pond after 5 minutes?

(c) Copy and complete this table.

Time in minutes	0	1	2	3	4	5	6	7
Volume of water in litres	2500							

(d) On graph paper, draw and label axes like these.
Plot the points from your table.

(e) Join the points you have plotted with a line and extend it.

(f) Use your graph to say how much water is left in the pond after 10 minutes.

(g) How much water is left in the pond after 12 minutes?

(h) After about how many minutes will there be 500 litres left in the pond?

(i) About how long will it take to empty the pond?

(j) The pumping started at 10:00 a.m.
At about what time will the pond be empty?

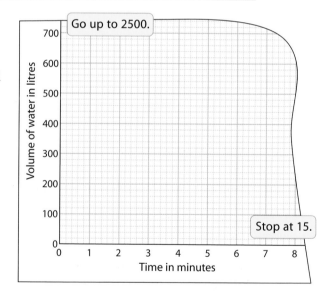

B Graphs and rules

1 Highrise Construction uses this rule to work out how many
days' holiday its employees are allowed each year.

number of days' holiday = (number of years employed ÷ 2) + 20

(a) How many days' holiday are allowed for Jon, who has been employed for 6 years?

(b) Copy and complete this table.

Number of years employed	0	2	4	6	8	10
Number of days' holiday	20					

(c) Plot the points from your table using axes like these.

(d) Join the points with a line and extend it.

(e) How many days' holiday would you get if you had been employed for

(i) 12 years   (ii) 16 years

(f) Joy gets 29 days' holiday.
How long has she been employed for?

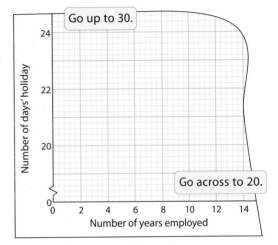

2 Speedyfix uses this formula to calculate the charge for a repair job.

$c = 10h + 35$

c is the charge in pounds and
h is the number of hours the job takes.

(a) Copy and complete this table showing the charge for different lengths of job.

Hours taken (h)	1	2	3	4	5
Charge in £ (c)		55			

(b) Draw and label axes like these.
Plot the points from your table.
Join them with a line and extend it.

(c) Use your graph to work out the cost of a job that takes $4\frac{1}{2}$ hours.

(d) A job costs £120.
How long did it take?

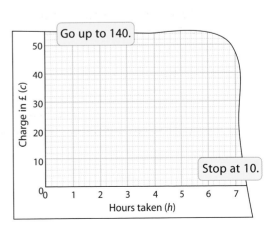

Mixed practice 6

1 How much is left when you use 100 millilitres of vinegar from a $\frac{1}{2}$ litre bottle?

2 This set of 'steps' is made from centimetre cubes.
   Each layer is a rectangle.

   Find the volume of the steps in cm³.

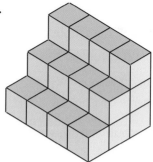

3 Work out each of these.

   (a) $7 \times (2 + 1)$ (b) $1 + 2 \times 7$ (c) $10 \div 2 + 3$ (d) $(9 - 7) \times (4 + 5)$

4 Digby's Drawers make cabinets.
   They use this rule to work out the price in pounds.

   $P = 30 \times d + 45$

   P is the price in pounds and d is the number of drawers.

   Use their rule to work out the price of a cabinet with

   (a) 2 drawers   (b) 6 drawers

5 Stamps cost 30p each.

   (a) How many stamps can Mia buy with £2?

   (b) How much money would she have left?

6 Find the value of $5(n - 3)$ when $n = 7$.

7 Change 1.2 litres to millilitres.

8 Calculate the area of this triangle.

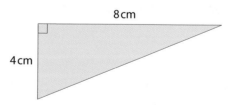

9 Fifi bought a pen and some postcards.
   The pen cost 55p and each postcard cost 40p.
   She spent £5.35 in total.
   How many cards did she buy?

35 Fractions, decimals and percentages

A Fraction and percentage equivalents

1 For each of these, write down

    (i) the fraction shaded   (ii) the percentage shaded

(a)

(b)

2 Sort these into four matching pairs.

3 Tina and Ron each have an identical bar of chocolate.
Tina eats $\frac{1}{4}$ of her bar. Ron eats 10% of his bar.
Who has eaten more chocolate? Explain how you know.

4 Write each of these lists in order, starting with the smallest.

(a) $\frac{1}{4}$, 100%, $\frac{3}{4}$, 10%   (b) 75%, $\frac{1}{4}$, $\frac{1}{10}$, 50%

B Fraction and decimal equivalents

1 Write down the decimal equivalent to each of these.

(a) $\frac{1}{10}$ (b) $\frac{3}{10}$ (c) $\frac{6}{10}$ (d) $\frac{9}{10}$

2 Sort these into four matching pairs.

$\frac{3}{4}$ $\frac{4}{10}$ $\frac{2}{10}$ $\frac{7}{10}$ 0.2   0.75   0.7   0.4

3 (a) A carton contains $\frac{1}{4}$ litre of juice. Write this as a decimal of a litre.

   (b) A bottle contains 0.5 litre of water. Write this as a fraction of a litre.

4 Which is bigger, 0.7 or $\frac{3}{4}$?
Explain your answer.

5 Write each of these lists in order, starting with the smallest.

(a) 0.25, $\frac{1}{10}$, 0.4, $\frac{1}{2}$ (b) $\frac{3}{4}$, 0.6, $\frac{7}{10}$, 0.9

c Fraction, decimal and percentage equivalents

1 Copy and complete this table.

Fraction		Decimal		Percentage
$\frac{3}{4}$	=		=	
	=	0.6	=	
$\frac{7}{10}$	=		=	
	=		=	40%
	=	0.9	=	

2

E	G	I	H	L	N	O	P	R	T	A
$\frac{1}{2}$	20%	0.3	$\frac{1}{4}$	$\frac{4}{10}$	0.7	$\frac{3}{4}$	80%	0.6	90%	0.1

Use this code to find a letter for each fraction, decimal or percentage below.

Rearrange each set of letters to spell an animal.

(a) 0.75, 30%, $\frac{7}{10}$, 0.4

(b) $\frac{2}{10}$, 0.9, 60%, $\frac{3}{10}$, 50%

(c) 75%, 0.25, 70%, $\frac{6}{10}$, 30%

(d) 0.5, 40%, $\frac{7}{10}$, 0.8, 25%, $\frac{1}{10}$, $\frac{9}{10}$, 50%

3 Which is closer to 30%, $\frac{1}{4}$ or 0.4?
Explain your answer.

D Percentage of a quantity, mentally

1 Work these out.

(a) 50% of 16   (b) 25% of 16   (c) 75% of 16

(d) 50% of 80   (e) 25% of 80   (f) 75% of 80

2 Work these out.

(a) 10% of 40   (b) 20% of 40   (c) 30% of 40   (d) 5% of 40

(e) 10% of 120   (f) 5% of 120   (g) 40% of 120   (h) 45% of 120

(i) 10% of 200   (j) 5% of 200   (k) 30% of 200   (l) 55% of 200

3 A car salesman gives 10% discount for a cash payment.
Jason buys a car costing £5400.
How much discount does he get for paying in cash?

4 20% of an orange's weight is peel.
If an orange weighs 80 g, how much does its peel weigh?

5 A packet of sweets normally contains 200 g.
It now contains '25% extra free'.
How many grams extra are in the packet?

36 Two-way tables

A Reading tables
level 3

1 Laura needs to travel from Birmingham to Glasgow.
The table shows the different ways she can travel.

Transport	Time	Distance	CO_2 emissions
Car	5 hours 25 minutes	301 miles	79.1 kg
Coach	7 hours 45 minutes	325 miles	46.6 kg
Plane	2 hours 5 minutes	284 miles	72.1 kg
Train	4 hours 50 minutes	275 miles	26.5 kg

(a) How long does it take to travel by car?

(b) Which means of transport is the quickest?

(c) Which means of transport goes the shortest distance?

(d) Which journey produces 72.1 kg of CO_2?

(e) Laura is concerned about the environmental effects of her journey.

    (i) Which means of transport has the lowest CO_2 emissions?

    (ii) Which means of transport has the highest CO_2 emissions?

B Distance tables
level 4

1 This table shows the distances in miles between some English cities.

Bristol					
171	Cambridge				
83	250	Exeter			
120	60	200	London		
74	100	154	56	Oxford	
76	131	112	80	66	Southampton

(a) How far is it between Exeter and London?

(b) How far is it between Cambridge and Oxford?

(c) Which two cities in this table are closest together?

(d) Dennis drives from Cambridge to Exeter.
He then drives from Exeter to Bristol.
How far has he driven altogether?

2 (a) Reshma drives from London to Bristol.
How far is this?

(b) On her return journey she drives from Bristol to Oxford and then from Oxford to London.
How far is this?

(c) How much further did she drive on the return journey?

37 Scale drawings

You need centimetre squared paper and an angle measurer in section A.

A Simple scales

1 This scale drawing of a wine label is on centimetre squared paper.
1 cm on the drawing represents 2 cm on the real label.

(a) What is the height of the real label?

(b) What is the width of the real label?

(c) Measure the length of any edge of the label on the scale drawing.

(d) What is the length of an edge on the real label?

2 This is a sketch of the end wall of a house.

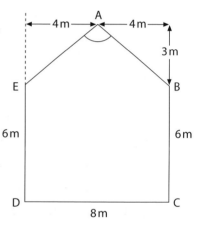

(a) On centimetre squared paper, make a scale drawing of the wall using a scale where 1 cm represents 2 metres.

(b) Measure the length AB on your scale drawing.

(c) How long will AB be on the actual wall?
Use the correct units in your answer.

(d) Measure the angle at A, marked on the sketch.

(e) What will this angle be on the real wall?

3 The triangle on centimetre squared paper is a scale drawing of the blue triangle.

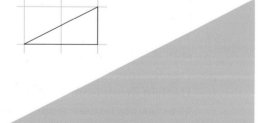

(a) By measuring the blue triangle, work out what scale has been used.

(b) Using the same scale, a drawing is done of a rectangle 12 cm long.
How long is the rectangle on the scale drawing?

(c) The rectangle is 2.5 cm high on the scale drawing.
What is the height of the real rectangle?

4 This is a sketch of a sail for a boat.

(a) Make a scale drawing of the sail using a scale where 1 cm represents 1 m.

(b) Measure the side BC on your scale drawing, to the nearest 0.1 cm.

(c) What is the length of side BC on the real sail?

(d) What is the angle at B?

(e) What is the angle at C?

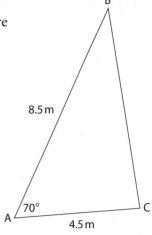

B Harder scales

1

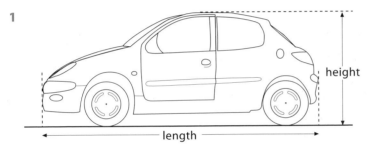

This car is drawn to a scale where 2 cm represents 1 metre.

(a) What does 1 cm on the drawing represent in real life?

(b) Measure the height of the car on the drawing.

(c) What is the height of the real car?

(d) Measure the length of the car on the drawing.

(e) What is the length of the real car?

(f) The car is to be redesigned so that it is 4 metres long. How long will it be when drawn to the same scale as the drawing above?

2 A bicycle is drawn to a scale where 5 cm represents 1 metre.

(a) The diameter of the wheels is half a metre. What would this be on the drawing?

(b) What does 1 cm on the drawing represent in real life?

(c) The top of the saddle is 80 cm from the ground. What distance would this be on the drawing?

(d) On the drawing, the distance between the wheel axles is shown as 6 cm. What is this distance on the real bike?

38 Using percentages

You need a pie chart scale.

A Percentage bars
B Interpreting pie charts

1 This percentage bar shows the nutritional content of a packet of crisps.

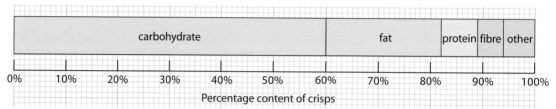

What percentage of the crisps is

(a) carbohydrate   (b) fat   (c) protein   (d) fibre

2 This pie chart shows the costs of keeping a car on the road.

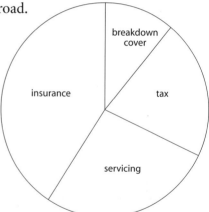

(a) What percentage of the cost of keeping a car is insurance?

(b) Is the cost of tax more or less than the cost of servicing?

(c) What percentage of the cost of keeping a car is breakdown cover?

(d) What percentage of the cost of keeping a car is tax?

3 This pie chart shows how the maths department spent its budget last year.

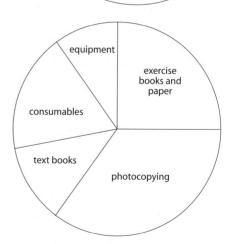

(a) What did the maths department spend the most money on?

(b) What percentage of the budget was spent on consumables?

(c) What percentage was spent on equipment?

(d) If the department budget was £8000, roughly how much was spent on exercise books and paper?

c Drawing pie charts

1 This information shows the contents of three different types of chocolate.

	White chocolate	Milk chocolate	Plain chocolate
Protein	6%	8%	4%
Carbohydrate	57%	57%	65%
Fat	34%	30%	30%
Water	3%	5%	1%

Draw a pie chart with radius 4 cm for each type of chocolate.

D Comparing

1 This chart shows percentages of male and female adults in Great Britain in 2002/2003.

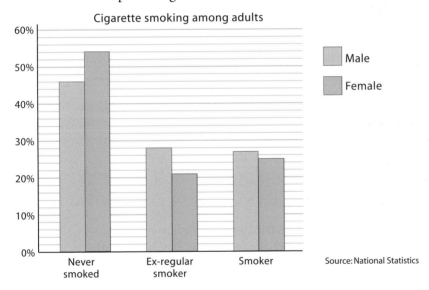

Source: National Statistics

(a) What percentage of males had never smoked?

(b) What percentage of females were smokers?

(c) Describe two differences between the males and females shown in this chart.

2 (a) These figures show the spending of two different charities.
Use this information to draw a bar chart like the one above, comparing the spending of the two charities.

	Emergencies	Health	Education	Fund-raising	Administration
Charity A	16%	29%	31%	22%	2%
Charity B	25%	38%	15%	14%	8%

(b) Describe the main differences between the spending of the two charities.

39 Conversion graphs

You need graph paper for section B.

A Using a conversion graph

1 The areas of floors and carpets are usually
measured in square metres or square feet.

This graph can be used to convert
between square feet and square metres.

(a) Use the graph to convert

    (i) 150 square feet to square metres

    (ii) 280 square feet to square metres

    (iii) 25 square metres to square feet

    (iv) 12 square metres to square feet

(b) The Balcony Café in the Crowne Plaza
Hotel in New York has a floor area
of 2000 square feet.

Use the graph to help you convert
this area to square metres.

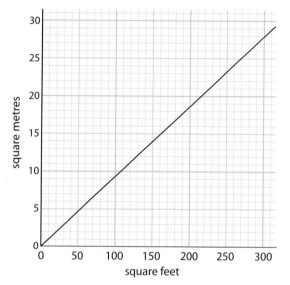

B Drawing a conversion graph

1 In older cookery books, liquids may be measured in fluid ounces.
This table shows some volumes in fluid ounces and in millilitres (ml).

Fluid ounces	0	5	10	15	20	25
Millilitres	0	140	280	420	560	700

(a) On graph paper draw and label
axes like these.
Draw a conversion graph.

(b) Use your graph to convert

    (i) 18 fluid ounces to ml

    (ii) 250 ml to fluid ounces

(c) I have a recipe for stuffed vine leaves
that uses $7\frac{1}{2}$ fluid ounces of yoghurt.
How many millilitres of yoghurt is this?

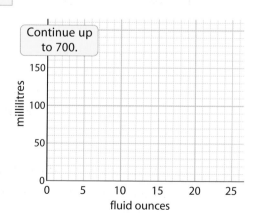

Continue up
to 700.

Mixed practice 7

You need a pie chart scale.

1 Write 75% as a fraction.

2 This is a scale drawing of a bedroom.
   The scale of the drawing is
   2 cm to 1 metre.

   (a) **(i)** Measure the length of the
           bedroom on the drawing.

       (ii) What is the length of the
            real bedroom?

   (b) What is the width of the
        real bedroom?

   (c) The bed is 2 m long.
        How long should it be
        on the scale drawing?

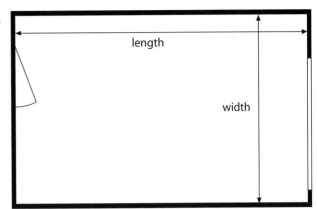

3 This pie chart shows the reasons people gave when they
   complained about adverts on television.

   (a) Use a pie chart scale to measure the percentage
        of people who complained because
        they thought an advert was misleading.

   (b) What percentage complained because they
        thought an advert was harmful?

   (c) There were around 8000 complaints altogether.
        Roughly how many were from people who
        thought an advert was offensive?

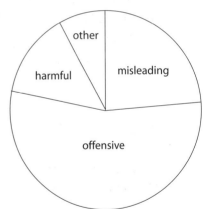

4 Work these out.

   (a) 10% of 60   **(b)** 20% of 60   **(c)** 20% of 80   **(d)** 25% of 60   **(e)** 10% of 500

5 A survey asked students whether they walk, are driven or get the bus to school.
   The results are in the table.

   (a) What percentage of students
        aged 5–10 get the bus to school?

   (b) Are these statements supported
        by the data?
        Write 'true' or 'false'.

Transport	Age 5–10 (%)	Age 11–16 (%)
Walk	49	44
Car/van	43	22
Bus	6	29

Source: National Statistics

   (i) About half of the students aged 11–16 get the bus to school.

   (ii) About half of the students aged 5–10 walk to school.

6 This diagram shows the side view of a desk.

(a) Using a scale of 1 cm to represent 10 cm, make an accurate scale drawing of the side view of the desk.

(b) What is the real length *f*?

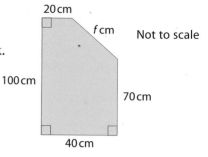

20 cm

f cm   Not to scale

100 cm

70 cm

40 cm

7 This chart shows some data about young people and smoking between 1990 and 2000.

Percentage of pupils aged 11–15 who regularly smoked cigarettes

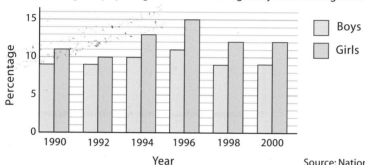

☐ Boys

☐ Girls

Source: National Statistics

(a) What percentage of the boys smoked regularly in 2000?

(b) In which of these years did the lowest percentage of girls smoke regularly?

8 In August 2007 Chloe went to South Africa on holiday.
She used this conversion graph for pounds (£) to rand, the currency of South Africa.

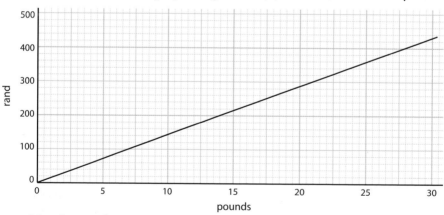

(a) Use the graph to

(i) change £5 to rand   (ii) change 200 rand into pounds

(b) Chloe bought a handbag on holiday that cost 360 rand.
About how much was this in pounds?

9 Write each of these lists in order, starting with the smallest.

(a) $\frac{3}{4}$, 0.3, 10%, 0.65   (b) 0.8, 25%, $\frac{9}{10}$, 0.75